Miniature Sheet of Patterns and price list of same mailed on receipt of 3-cent stamp.

SHIPMAN'S MANUAL OF 87 ASSORTED PATTERNS, 60 CENTS.

A selection of desirable designs, Six Alphabets, also an Elaborate Treatise on Bracket Sawing, Inlaying, Preparation of Woods, Polishing &c.

PRICE LIST OF SHIPMAN'S NEW DESIGNS.

No.	NAMES.	Size.	Price of Pattern
50	Paper Holder	10x13	10c.
51	Photograph Frame	7x10½	5
52	Lambrequin Bracket	5x 7	10
53	Lady's Work Box	4x8x12	10
54	Photograph Frame	8x10½	5
55	Shrine Bracket	6x16	10
56	Large Photo. Frame	10x13	10
57	Lyre Bracket	7x11	5
58	Hanging Basket	13x17	10
59	Fancy Centre Piece	6x13	5
60	Easel	4x 6	5
61	Cigar Stand	5½x 8	10
62	Large Bracket	13x15	10
63	Photograph Frame	8½x12	10
64	Pen Rack	4x 4	5
66	Bracket	13x14	10
67	Photograph Frame	8x12	10
68	" "	4½x 7	5
69	Corner Bracket	6x15	5
70	Photograph Frame	5x 8	5
71	Corner Bracket	6½x 9	5
72	Photograph Frame	9x15	10
73	" "	9x12	10
74	Cross	10x15	10
75	Bracket with L. Glass	8½x13	10
76	Bracket	13x16	10
77	Easel	7½x12	5
78	Flower Pot Holder	7x 9½	5
79	Glove Box	4x4x12	10
80	Card Basket	7½x 8½	5
81	Photo. Frame and Doors	8x13	10
82	Two-Shelf Bracket	9x13	10
83	Hair Receiver	6x 8	5
84	Masonic Bracket	6x10	5
85	Bracket	10½x12	5
86	Photograph Frame	7x11½	5
87	Clock Shelf	7x16	10
88	Fancy Piece	4½x 6	5
89	Clock Shelf	5x16	10
90	Hanging Basket	10x14	10
91	Photo. Frame with doors,	10x11	10
92	Easel	7x12	5
93	Thermometer Holder	3x14	5
94	Easel	6x14	5
95	Corner Bracket	6x18	5
96	Photograph Frame	8x12	5
97	Thermometer Holder	3x13	5
98	Card Receiver with pillar and base turned	7x10	10
99	Photograph Frame	6x 8	5
100	Match Safe	4x 6	5
101	Photograph Frame	5x 7	5
102	Bracket	10x11	5
103	Slipper Case	11x15	15
104	Photo. Frame	9x14	10
105	Card Basket	6x 6	5
106	Towel Rack	12x16	10
107	Photo. Frame	12x15	10
108	Wall Pocket		15
109	Table Mat	9x12	5
110	Card Basket	5x 9	15
111	Knife Box	4x12	15
112	Hanging Clock	7x18	15
113	Corner Bracket	10x16	5
114	Cross	8x15	10

No.	NAMES.	Size.	Price of Pattern
115	Motto	9x12	10c.
116	Card Basket	6x 8	15
117	Brush and Comb Case	12x19	10
118	Silhouette	4x 7	5
119	Easel	9x16	10
120	Comb and Brush Case	8x14	10
121	Shrine Bracket	9x17	10
122	Card Basket	4x 7	10
123	Photo. Frame	6x 8	5
124	Pen Rack	6x 6	10
125	Table Mat	9x12	5
126	Wall Pocket	14x16	15
127	" "		10
128	Hanging Basket	9x13	10
129	Bracket	11x14	10
130	"	11x15	10
131	Match Safe	4x 7	10
132	Corner Bracket	6x 9	5
133	" "	6x15	10
134	Easel	4x 7	5
135	Three-Photo. Frame		10
136	Silhouette	2x 5	5
137	"	2x 9	5
138	"	3x 4	5
139	"	3x 6	5
140	"	3x 4	5
141	"	2x 7	5
142	"	4x 7	5
143	"	3x 3	5
144	Motto	5x14	10
145	Silhouette	3x 3	5
146	Card Basket	8x 8	5
147	" "	8x 8	5
148	Sign	7x18	10
149	Towel Rack	8x18	10
150	Corner Bracket	7x 9	5
151	Card Basket	8x 8	5
152	Hand Mirror Frame	5x11	5
153	Thermometer Stand	3x15	5
154	Slipper Case	5x11	10
155	Motto	11x12	10
	Lord's Prayer, No. 2	15x18	25
156	Italian Pendulum Clock	12x13	15
157	Mantle Clock, striking	10x13	15
158	Hanging Clock	11x18	15
159	Cross	[illegible]x12	5
160	Hanging Basket	11x16	10
161	Spool Stand	4x 6	10
163	Table Mat	9½x13	5
167	Silhouette	4x 4	5
168	Cupids	2x 8	5
169	"	3½x 3½	5
170	"	4½x 5	5
171	Silhouettes	3x 6	5
172	"	4x 5	5
173	"	4x 5	5
174	Vase	5½x 4	5
175	Cupids	6x 7	5
176	Horse Shoe	6x 6½	10
177	Cupids	6x 7	5
178	Silhouettes	5x 6	5
179	"	2x 4	5
186	Motto	6x18	10
187	"	6x15	10

PATTERNS FOR WINDOW DECORATIONS ON PAGE 17.

THIS LOT COMPLETE, $1.00.

Letter.	NAMES.	Size.	Price.
a	Lambrequin Bracket	12x39	25c.
b	Turned and Sawed Swinging Bracket	8x13	10
c	" " " " "	7x10	10
d	" " " " "	10x11	10
e	" " " " "	7x10	10
f	Hanging Flower Pot Basket	9x13	10
g	Fancy Turned Stand	14x27	25
h	Window Garden	9x31	25

Foreword

This reprint of the 1881 A. H. Shipman Bracket Saw Company catalog illustrates the founder's merchandising efforts. It was an early example of a business that aggressively targeted a niche market, in this case amateur woodworkers and, especially, children, through advertising techniques that were advanced for the time. The catalog speaks to parents suggesting the fret-sawing hobby as marketable and profit-making skill and as a way to develop a good work ethic.

The Shipman & Binder Manufacturing Company started out in a small way in 1875 at 111 Water Street on the east side of the riverside industrial district of Rochester. Born in 1850, Albert H. Shipman spent his boyhood days in Syracuse, New York. Shortly after his marriage in Wayne County in 1873, he moved to Rochester. The only Binder on record, and Shipman's likely partner in 1875, was a Julius Binder. Born in Germany in 1816, he was recorded in Rochester as a piano maker in the 1880s. In 1876, the year of America's hundredth anniversary, the company introduced the "Centennial Bracket Saw" in that year's catalog, a pocket-sized, one-fold, yellow paper leaflet. The no. 1 model ($2.50) was a wooden, table-mounted, hand-crank mechanical fret saw, and the no. 2 model ($4.50) was a wooden, three-legged, foot-powered machine. These were light-duty amateur saws intended for hobbyist and children. "Parlor brackets, mottoes, and ornamental scroll work of every description" were the intended output. Some extra saw blades and a selection of pattern designs were included to inspire the user.

The offering must have been a quick success judging from the 1878 catalog. The "New and Greatly Improved Centennial Bracket Saw" was featured on the cover along with an illustration of a fashionably dressed and coifed lady at work on the foot-powered contraption. New items included a lathe attachment, with a gouge and skew chisel, to tempt the mechanically curious as well as expand the product to a multi-faceted application—all for $8. The basic saws were still available at the same price as in 1875, but now a variety of combinations could be ordered, including a drilling attachment and some accessories, such as a pantograph and a saw-filing clamp to re-sharpen your jig saw blades. Rounding out the offerings was a professional grade "Trio or Complete Workshop," a workbench, lathe, jigsaw, and bench vise, all in one unit, for $15.

The 1878 catalog was a giant leap in merchandising effort over the modest first edition. With the lady fret sawyer on the front cover, the simplicity and

i

universal application of the product was emphasized, and options were expanded within the catalog to accommodate many levels of interest. A more substantial version of the foot-powered saw and work table was introduced and targeted to the independent craftsman.

The year 1878 marked a crossroads for Shipman & Binder. The company sustained $18,000 in damage from a fire that started just after the workday ended on November 26. The Hydraulic Building on Water Street, where the saw factory was located, was one of many in this industrial district that derived water power from raceways cut out along the upper banks of the Genesee River adjacent to the water falls. The Hydraulic Building was near the ninety-foot High Falls. Supposedly, friction on a belt-driven machine in the saw factory on the first-story started a fire that followed up the drive belt openings in the four-story building. The other business occupants included the Rochester Hydraulic Company (the provider of power to the occupants), four boot and shoe manufacturers (including one who reported the loss of many tool kits valued at $25 each), a shoe tool manufactory (Townsend & Wiseman), a machinist, a dental manufactory, and a ladies undergarment maker. The upper story occupants sustained total loss. By the following year, the company, now named the A. H. Shipman Bracket Saw Company, had relocated to nearby number 7 Andrews Street.

This catalog, an expansion of the 1878 version, introduced Victorian-style images of the factory and the products. The back cover features an illustration of a contented family of five seated in the parlor. Four members are occupying themselves in various pursuits, while the young boy of the family is hard at work on his cast iron, $3.00 Holly Scroll Saw (page 16). Cast iron machines had been introduced shortly after the move to Andrews Street. The star product was the Prize Demas machine (page 18), a larger, beefier, more versatile version of the Holly. It was basically a foot-powered lathe that with some attachments would convert to a jigsaw. The effort to attract youth to the products is illustrated by a contest (page 2) with $1,000 prize "to be given to the boys who do the nicest work upon a scroll saw." (There evidently were different publicity efforts around the country as well. A Holly Scroll Saw purchased in Vermont in the 1990s included a story about the saw, which had been won in a scroll sawing contest by a boy in the late-nineteenth century.) The Shipman Engine (page 23), a bench-side steam engine that was small enough to be portable and fulfill the needs of a one-man shop, was targeted to the professional workshop owner. The engine later developed into the main focus of the business.

In its early days the company had employed a ten-man crew, but by the time the company moved to a new location at Bismark Place in 1880, it required a workforce of two hundred to meet demand. In 1885, the company name changed

continued at back of catalog

again to the Shipman Engine Company, emphasizing its small steam engine. In the same year, the Shipman Engine Company of Boston was established.

Albert Shipman first included a history of the scroll saw in the 1880 catalog in which he claims to have introduced the inexpensive amateur scroll saw to the world in 1876. The history in this 1881 catalog tells the charming story of the American lady who observed fret sawing in Switzerland (one can imagine using a fret saw to make a fancy Black Forest clock) and brought back the practice to America (page 3). Shipman admitted that the publisher of *Youth's Companion* magazine, Perry Mason & Co. of Boston, was the original developer of the amateur saw outfit. Shipman's acknowledgment, along with a trade card from Mason advertising a Holly saw, confirms the existence of a business relationship between A. H. Shipman Bracket Saw Company and Perry Mason & Co. Shipman had been careful to claim only to have introduced inexpensive, or "cheap," amateur machinery. A circa 1879 catalog from the W. F. Barnes Company of Rockford, Illinois, which was founded in 1871, shows amateur machines in the twelve-dollar price range. The Holly cost only three dollars.

Shipman catalogs of the 1880s, including this one, offered a block plane that looks much like the Stanley no. 101 (page 26). There are also little, hand-cranked, cast iron drills (page 28) and six-piece, short carving chisel sets (page 27), also seen in Millers Falls catalogs. The Holly machine illustrated in this catalog (page 16) is very hard to distinguish from the Millers Falls "Companion Scroll Saw." These similarities suggest that either a close relationship had existed between the two companies or copying of one or the other's design had occurred.

The premature death on July 30, 1888, of the thirty-eight-year-old Albert Shipman was announced in the *Rochester Union Advertiser*. His obituary described his recent history of lung and throat difficulties. He had sought treatment in nearby Geneva, New York, where he expired.

In 1893 the factory was moved across the river to the Brown's Race industrial section, the steam engine business was expanded, and all mention of scroll saws was dropped from the directories. The last business directory entry for Shipman was in 1900. Without the dynamic Albert Shipman to guide it, the company had experienced a drawn-out, slow death.

In this catalog Shipman addressed the parents of the day. His message was that a child with a fret saw could learn useful mechanical skills as well as the rewards of hard work. Achieving these commendable goals continues to challenge parents 120 years later.

Frank Kosmerl
Rochester, N.Y.
May 2002

This reprint was published in conjunction with the 2002 Annual Meeting of the Early American Industries Association at Rochester, New York. The purpose of the EAIA is to encourage the study of and better understanding of early American industries in the home, in the ship, on the farm, and on the sea. Membership in the EAIA is open to any person or organization sharing its interests and purposes. For membership information, write to Elton W. Hall, Executive Director, 167 Bakerville Road, South Dartmouth, MA 02748, E-mail <EAIA@ fastdial.net> or visit the EAIA Web page at www.EAIAinfo.org.

Thank you to EAIA member Dan Semel for providing the A. H. Shipman catalog.

PREFACE.

I N offering my new and enlarged Catalogue to the public, I feel that a few words of explanation regarding the introduction and continued developments of the FOOT POWER SCROLL SAW and TURNING LATHE, for the amateur mechanic, would be of great benefit to those who contemplate the purchase of either or both. Five years ago such a thing was comparatively unknown; but the introduction by me of the then popular Centennial Bracket Saw set the ball a rolling. Since then it has been almost impossible to manufacture them sufficiently fast to meet the increasing demand. The improvements over the old, and the new attachments I have each year added, are certainly wonderful; and now the **Prize Demas** and **Holly** stand pre-eminently at the head of all amateur tools. The reasons for this are numerous. I was the first to introduce them, and at the start determined to make this a specialty; to do this and be successful, I found it necessary to please my customers and give them satisfaction. Each purchaser was kindly asked to write me if his machine was not perfectly satisfactory. In this manner I found out any weak points or improvements that might assist in making my machines popular to the general public. My experience in the past five years, with the assistance I have received from my large list of patrons, has given me advantages which none of my imitators possess; and while a number of inferior articles are forced upon the market by these imitators, it requires but one to decide their further introduction in the same neighborhood. But even this should not be, for a *good Scroll Saw* is to every family what a good teacher is to a school; therefore everything possible should be done to increase its introduction.

Since the first introduction of these machines there has been a steady and rapid demand for neat, practical and artistic designs, and now the very best talent is employed to produce not only articles for home adornment, but such as are of practical use in every household, and in such a manner as will enable the amateur to closely imitate, and in a majority of cases to excel those purchased at enormous prices of dealers. The demand of the amateur became so great that it was found necessary to issue a publication devoted exclusively to their interests. In it, each month, we so illustrate and describe new and original designs as will enable the most inexperienced to produce the finest results, and we feel safe in saying that in the next five years its success will be such as will place it among the leading publications of the country.

As before stated, I make this business a specialty, and it is the only one of the kind in the world. My factory contains 17,103 square feet of working room, all of which is devoted exclusively to amateurs' tools. Goods from here are shipped to every known part of the globe. With the assistance of the public, I have each year very largely increased the interest and sales of these goods, and hope this year, with my increased facilities and extra inducements offered, that I shall be able to more than double my former endeavors.

Most respectfully,

A. H. SHIPMAN,

Rochester, N. Y.

FACTS THAT TELL.

"I say, Jack, this is Bill Allen, didn't know him, did you?"

"Why no, is that you, Bill; where on 'arth you been, eh, old boy? you look's though a streak of luck struck you; eh, Dan, don't he?"

"Waal I should smile; what on 'arth's the matter with him,—going to college I guess; eh, boys, looks it, don't he?"

"I am surprised and yet glad to think there is such a change in me; now listen and I will tell you how it occurred, and how you can accomplish the same result. You remember little Johnnie Goodyear, only 13 years old; you remember my telling you about his getting a Holly Saw, and what nice brackets, picture frames, and such things he made with it; you remember I told you how much money he earned one Saturday; more than any of us had earned in a whole week. Well, one day I just said to myself: 'Here am I, a great boy 15 years old, hanging around corners, blacking other folks' shoes, and can't even read my own name, while there's little Johnnie going to school, already a fair scholar, and besides this earning more money every week than I.' I could not stand this, so I just pitched in, got every job I could, saved every penny until I got enough to buy me a Holly Saw. That was five weeks ago, and here I am as you see me; have plenty to live well, and go to school every day; have just heard of those big prizes and am going down to investigate the matter; for I am bound to have one. Your chances are as good as mine, and should you not succeed in getting a prize the fact that it has been the means of your getting a Holly Saw, with which you are making lots of money, will be a prize equal to thousands of dollars in your future lives. There, that's right, there goes Ike Spaulding on a keen jump off for a Prize Holly Saw. I wish him every success; and now boys, take my advice and do likewise. Good-by; hope to meet you when prizes are distributed."

☞ These facts, in one shape and another, are making themselves heard from every section of the country, and parents should see to it that their children are not the last to be encouraged. Order a machine at once, that they may have time to learn and get one of those prizes which will be awarded May 1st., 1882.

FRET-SAWING;

PAST, PRESENT AND FUTURE.

CROLL or Fret Sawing of fancy articles originated in Switzerland, where, to this day, in many hamlets, it is their whole support, using, in all cases, the old style hand frame. A Boston lady, while traveling there, conceived the idea of introducing this most fascinating work into this country as a novel and useful mode of recreation. On her return she brought with her a number of these small frames, with a quantity of fine blades and patterns. In connection with her other business, she introduced this charming work, and found it gave great satisfaction, and that applicants for these outfits were so numerous that she had often to renew her stock from abroad. But the interest grew so rapidly that she commenced the manufacture of the frames in this country. Up to this time the cost of these outfits and instructions was quite an item, and naturally the wealthy alone could enjoy them. But the enterprising firm of Perry Mason & Co., publishers of the *Youth's Companion*, saw the necessity of placing an article of this kind upon the market that would reach the masses, and began the manufacture and introduction of their cheap outfit, which gave good satisfaction and had wonderful large sales, reaching into the the the hundreds of thousands.

This continued for several years, when I, all unknown to any of the above facts, produced the Centennial foot-power Scroll Saw, which I found to be an article that the lovers of Fret Sawing had been longing for ; something cheap, yet practical, that could be driven by foot power. This fact I learned of Perry Mason & Co., upon my first presentation to them of this most popular machine. This is now five years ago. Since then many changes and improvements have occurred. The old clumsy wood frame machine has given place to the neat and tasty iron frame. The then simple Scroll Saw is now giving place to the combined Scroll Saw and Lathe. When in the past this work was most all done for pleasure and amusement, it is now fast becoming a part and parcel of practical earnings of every-day life. Boys now indulge in it not only to satisfy their great desires for tinkering with machinery, but also to earn and lay up money ; hundreds of letters bear testimony to these facts.

Fathers find in these machines that which enables them to spend a great many pleasant hours at home, earning a few extra dollars, at the same time decorating their homes with those articles of beauty which help to make it pleasant to themselves and children; also educating the young folks in those arts which ennobles the soul and creates a desire for such things as will make true men and women of them ; he will also find that as they grow up their interest in the work executed on said machine has grown with them, and the moment they are old enough they will improve each spare moment until they have accomplished their desired object or find that they were never intended for mechanics.

My observations in the past five years, I think, warrant me to say that the time is fast approaching, and is already far advanced, when this will be the universal method of

(4)

testing children to show who can and cannot be mechanics; and what a blessing it will be, for how many boys are there who are brow-beaten around by some foreman because he can never make a mechanic; with this machine this is overcome, and at home in his own time and pleasure, he decides for himself whether he can ever become a mechanic or not.

One of the greatest questions of the day that America has to contend with is, what shall be done with the children? One argues, and the laws are in his favor, that they must go to school until a certain age, and that manufacturers must not hire them under a certain penalty. While this is as it should be, another, who takes the side of the poor man, has equally as good an argument, and one that cannot be set aside. He takes for example the common laboring man, or some mechanic who, from some injury, can not earn his former wages; either of these, and many thousands there are, who have large families; it is impossible for him to support them; what can he do, except as they grow old enough, to set them to work at such labor and wages as they can get. This is done at the sacrifice of their education, and in the majority of cases these children grow up to be common laborers, when, if they could have had the chances of others, they would have made some of our best mechanics or professional men. While this is very hard, the question is, how can it be helped? One thing is certain, their living they must have; how to give them this, and still allow the children time to go to school, is the great point to consider.

The introduction of mechanical tools adapted especially for the young folks, and at their own fireside, has already and will continue to solve this problem more practically than anything that ever has or can be introduced. The simple Scroll Saw has already made many homes rejoice over its wonderful results. Many boys, during their leisure hours after school, have earned much more than they could have done had they gone to some factory to work; some of these have gone direct from this simple machine into large factories and received $2.50 per day, all on account of their skillful manipulation of this little saw.

These facts have been so fully demonstrated in the past five years that this simple little Scroll Saw has been so improved and added to that now, in its combined form, it enables the youth to educate himself to any trade; and the best part of this is, while he is doing this he still goes to school and receives that education which is so necessary to all, he at the same time earning more toward the support of the family than though he had been placed in some factory at a very low sum, where he may never have received a mechanical education, but have learned much of the disgusting habits of low society.

To illustrate the value of these machines, we give in detail some articles made on them, their cost value and profit:

			Cost		Worth	Time Sawing.	Profit.
Material for ½ doz.	Horse Shoes,	- - -	$0.30,	worth	$1.50	1 hr.	$1.20
"	Small Easels, - - -		0.50,	"	1.80	1½	1.30
"	Picture Frames, - - - -		0.50,	"	1.80	2	1.30
"	Mottoes, "God Bless Our Home,"		1.00,	"	6.00	5	5.00
" ¼ doz.	Lord's Prayers,	- - -	3.75,	"	15.00	20	11.75

These are a few of the thousands of articles made on these machines, which find ready sale; this all who have ever used them will testify to. Time for sawing varies according to skill of operator; time stated above is an average. Brackets manufactured in large factories are made by boys, who are sacrificing their education to help support their families. Why not give them a saw and let them make them at home, before and after school-hours, thereby receiving their education and still earning more money than when at the factory. Let every parent consider these points.

IN ORDERING GOODS

THE FOLLOWING RULES MUST BE STRICTLY OBSERVED, OTHER-WISE MISTAKES WILL NOT BE CORRECTED.

WRITE your Name, Post Office, County and State Plainly, at the bottom of each and every letter. If you have written us and refer to it in your letter, you must give us the date of your former letter, or we cannot refer to it. Foreign correspondents will please enclose envelope properly directed, as their letters are directed so differently from ours we are often unable to reply for want of proper directions. State exactly what you want and price of each article. Send your money by Post Office order or registered letter; you will then know if we receive it. We do not wish to send goods C. O. D. If you are afraid to trust us, give the money to your Express Agent, and tell him to send for the goods and return them to you. We send small packages by mail, but we cannot guarantee their safe arrival. To guarantee safe delivery send 10 cents to have package registered. If you wish to know what the express or freight charges will be, ask your Freight or Express Agent ; tell him what the goods will weigh, and he can tell you. We cannot always find out.

We appoint no special agents nor give any exclusive territory. If parties wish to canvass for and sell our goods, we will give them terms of discount. On no consideration will we send our goods on commission. Our goods are so generally diffused throughout the world that any one can satisfy themselves, by a little inquiry, as to their quality and our reputation for fair dealing. If any one will study up the history of the Bracket Saw, they will find that our first machine, introduced in 1876 and exhibited at the Centennial, received a world-wide reputation from the fact that for the first time a practical foot-power jig saw would be sold for less than $20 or $25. We placed our Centennial wood frame foot-power saw on the market at that time for $4.50, and every-one thought it was a marvel of cheapness. From that time to the present we have spent exclusively in perfecting our machines, until the Prize Holly and Demas saws are the result. We have kept the prices of our machines at the very lowest margin of profit, and have given our customers the benefit of our inventions, which, we are happy to say, they have fully appreciated by their enlarged orders.

We copy from no one, but use our own brains, and the public will find, by a little research or notice, that all new styles in low priced machines, first originate with us, and that other parties attempt to copy ours as near as they dare and then lay claim to be the original inventors.

READ THIS SPECIAL NOTICE.

For the greater safety of delivery; and for the certainty of giving perfect satisfaction, I advise all to order their machine shipped all set up. The freight will cost you no more, and then when you go to get it from freight or express company, you can see that your machine is all there and in good condition; if not so, you then can make express or freight company pay for or replace any parts broken. The greatest advantage you derive from so ordering is that you know your machine has been set up and tried before leaving factory, and consequently must be O. K. For setting up Prize Holly or Prize Demas, 50 cts. extra ; Nos. 4 or 5 Demas, Mechanics' Saws, $1.00 extra.

Being personally acquainted with Mr. A. H. SHIPMAN, and also with the extent and his manner of conducting business, I can fully guarantee his ability to fill all orders sent him, and his financial standing makes it safe for any parties sending him money for goods. To any one who may hesitate to send money direct to him, can send it to the American Express Office, in this city, and I will purchase the goods and forward them without extra charge.   J. B. PRENTISS,
ROCHESTER, N. Y.   *Agent American Express Company*

Read the Following Testimonials,

From Clinton, Iowa :

The Demas has arrived and I am perfectly satisfied with it. I have five boys ; they think it a thing of beauty and a joy forever. There has been no trouble about getting them out of bed in the morning, since its arrival. I cheerfully add my commendations to those of others, as to its excellence ; and as a machinist of fifteen years experience, do not hesitate to say it is more and better machinery for the price than I ever saw before.

I. W. BUNTLINGER, Manager,
Union Iron Works.

A. J. PIFFET, of New Orleans, La., writes that he has made $500 in the past year, besides his regular work.

From Chester, Pa.:

My little son and daughter are delighted with their saw, and in fact the household arrangements seem incomplete without one.   MRS. S. M. KEEL.

WASHINGTON HEIGHTS, ILL.

That beautiful Holly Saw has arrived. It is a perfect gem. In behalf of its owner, who is perfectly delighted, we return our thanks. No one but a lame boy can so well appreciate this fine machine. To his great delight he can run the treadle with *one foot.* Again let us thank you.
M. J. & K. L. LIVITT.

From Columbus, Ohio :

The Demas came all right ; am very much pleased with it. I have made some beautiful brackets, frames, &c. I received the saw in the evening, set it up myself, and made a nice bracket the same night. I am 13 years old.

WILLIE CONVERSE,
290 West 3d Ave.,

From Rising Sun, Md. :

Saw come all right, some time ago, and gives satisfaction in every respect.
E. H. KIRK.

From Alton, Ill.:

I received my saw in good shape ; it is a perfect little beauty, and can't be beat for the price.   D. W. BEAZNELL.

From Schenectady, N. Y.:

We received the Holly last Friday P. M. all right ; we set it up Saturday. I need not tell you that our boys are wild with delight over it. My husband is a genuine machinist, and thinks it a marvel of cheapness. MRS. E. D. CHAMBERS.

From Towanda, N. Y.:

About a week ago I bought one of your Scroll Saws, and like it better than a $20 saw I had.   WILLIE H. PATTERSON.

From Covington, Ohio :

I received, last week, the Holly Scroll Saw you sent me through the *Farm & Fireside,* of Springfield, Ohio. I am much pleased with it. This is the third saw I have used, and the Holly makes less noise and runs lighter than any I have tried ; it cannot be too highly recommended.
CLARENCE ALBAUGH.

From Alberton Md.:

The Holly Scroll Saw is not a toy, but a substantial machine which no man need to be ashamed to own or use.
W. B. GAMBRILL, M. D.

From Waverly, N. Y.:

I bought one of your saws last winter, and would not sell it for double its price ; it is a perfect gem. I sawed black walnut in pieces $\frac{1}{8}$ inch thick, together, making a thickness of $1\frac{3}{4}$ inches.
GEO. E. HAIR, Box 31.

From Halifax, N. S.:

Your Holly is a marvel for cheapness and utility, and needs only to be seen to be appreciated ; you claim nothing for it that it does not fulfill. There are several in use in Halifax and all hold the same opinion that I do.

FRED. B. WOODILE,
Morris St., West.

My machine is a little jewel. I have done some handsome inlaying and sawing, and have turned out several goblets on the Lathe, which the best turner in town pronounces excellent work.
CHAS. K. ROBB.

Abilena, Kan., Jan 1, 1881.

I have had my Demas Lathe and Saw about seven weeks. I made $10.00 worth before Christmas ; since then I have made as much more. The machine is a wonder to the town. Articles are in demand faster than I can make them.
F. S. HAFFORD.

Burgoon, Ohio, Jan. 27, 1881.

I received my Demas machine Dec. 31. I must say I am well pleased with it, for it is just such a machine as I have been wanting. I would not take $50.00 for mine if I could not get another.
H. G. HOGENDOBLER,

Villa Ridge, Ill., Jan. 11, 1881.

Manual of Fret Sawing and Turning,

WITH ILLUSTRATIONS.

Caution.—The amateur is frequently in too great haste to "make a bracket," and does not give sufficient time for practice; especially is this the case with the younger ones, and for this reason many get discouraged. The true way to be successful is to follow instructions explicitly. *No one should attempt to do a nice piece of work until they can saw on a line, circle, or cut an angle with ease and precision;* for this practice old cigar boxes are the cheapest and handiest. After you have learned to follow straight and curved lines, the next and only difficult point is to learn to turn a square or sharp corner, as in lesson I.

Lesson I.—Commence at one end and saw up to sharp point; now, without stopping the motion of the saw, you want to swing the piece of wood around, using the saw blade as a fulcrum for center; when you get so you can successfully do this you will find it of great value in executing work rapidly and nicely. At first you will find this a little difficult, for the reason you do not turn on the actual center of saw blade; this is caused by your pressing the wood forward slightly while you are turning it; now it should not be pressed in any direction, but if any way it should be held back a little, for the reason that it would then hit back of saw blade, which cannot cut; but this will be overcome by practice.

Lesson II.—This is a different practice, but will require no special instruction. Carefully follow the lines. Do not crowd or hurry your work. In case the saw works hard, occasionally apply a little soap or beeswax to the back of blade.

Lesson III.—This is a combination of the line, curve and angle, but differing from previous lessons in this respect: in preceding figures our object was to preserve outside line; in this we preserve inside. First drill small hole at *a*; unscrew upper saw fastener, insert blade through hole. Now proceed to cut out the design as already described.

Having practiced on lessons until you can saw true, either on line, curve or angle, you will be ready to advance a step in this fascinating art.

Lesson IV.—MAKING A BRACKET.—Having selected a good piece of black walnut about three-sixteenths of an inch thick, apply to it the design. There are several methods for doing this. The simplest is to paste design directly upon the wood, using flour paste. After sawing is done, the paper can be removed by moistening with water, but it is much better to be careful in putting paste on, and only put it on the part that you throw away. You can also secure design to the wood with small tacks, driven into parts of the wood which are to come away. Another good way: Procure sheet of "impression paper," lay paper on the wood, place design over it; take an instrument with *fine, hard point*, and trace around design. On removing impression paper, design will be seen neatly copied on wood.

It is better to cut away inside work first. Drill holes in every part which requires cutting away; this done, place wood, with design upwards, on the saw table. The inside of design being complete, remove outside of design in same manner. Having finished the various parts of the bracket, it is ready for finishing.

Smoothing off Work.—Take small half round file, and file corners true, and straighten all edges. Take sand-paper, rub bracket carefully. Under edges will be found ragged, but sand-paper will make them smooth. When a number of thicknesses are sawed, this is obviated.

Putting Work Together.—Small brads or screws can be used, if the bracket is quite small glue alone is sufficient. In case bracket is intended to carry some weight, use screws.

Oil—For oiling, boiled linseed oil should be used. Apply to wood, and when it is absorbed, rub over with a stiff brush or soft paper.

Shellac.—Take half-pint bottle of alcohol, fill about one-quarter full of bleached shellac in small pieces. After standing several hours this will be dissolved and ready for use. Apply to the wood with fine sponge or cotton. It dries so very quickly that several coats can be applied in a few moments.

Varnish is frequently used; does not give the wood as pretty an appearance as oil or shellac. If used, must be applied lightly and evenly. There are several kinds, varying in color; must be used according to color of wood.

Polish.—To polish well requires practice, care and patience, and would not advise amateurs to undertake the work unless very desirous of so doing. Prepared French polish is generally for sale at paint stores. In first place see that wood is smooth. Use fine sand-paper, and be sure to remove every scratch. Having obtained the polish,—light or dark, according to color of wood,—soak small bit of tow or cotton wool in the polish, and apply *evenly* to wood; add more polish to cotton wool, but before applying to wood place it inside a piece of linen rag, on which put drop or two of sweet oil—this prevents rag from sticking; rub wood again, giving circular motion to rag; repeat supply of polish and oil as required, until surface is uniformly polished.

Marquetry or Inlaying is a fascinating part of scroll sawing, and destined to become more popular. At first thought it may seem difficult to inlay one piece of wood into another, but the process is simple when you know how.

For first lesson we will take clover leaf, and inlay black walnut into white holly.

Fig. 5.   Fig. 6.

Take two pieces of wood, each one-eighth of an inch thick, walnut and holly; fasten them together either with common shoe pegs or screws. Let the walnut remain at top; secure the design to wood; drill small hole for saw blade, same as in fret-sawing, in which insert blade (No. 0); saw with your work on right of saw blade. Cut (Fig. 6) shows two pieces of wood, and angle at which blade should cut. From picture you will readily see how dark piece of wood will drop down and fit into light. If bevel is just right it will leave work smooth with no gaps. Having cut out clover leaf you can secure it in its place. Glue around edges of leaf, insert quickly in holly, let harden under pressure. Finish off work with sand-paper.

☞**How to use the Holly and Demas Saws.**—Amateurs should first learn to operate treadle, so they can run machine and talk at same time; even write and run saw. Having accomplished this, take piece of cigar box or other thin board, make straight and curved lines upon it (do not at first turn round). When you have made marks, place board close to saw, as near mark as possible, with hands on top of board; press down gently—not hard, but always down and forward at an even speed, not by fits and starts. Keep hands as near saw as possible. Always use thin lumber first, and saw slow; as you learn to saw you can learn to use machine to fullest capacity. To saw bracket or other piece of work, always place pattern on wood so grain will run lengthwise of weaker parts.

Overlaying.—When amateur has become master of his saw, so he can saw delicate and intricate work, he should do overlaid work, as this is very neat and a change. Ordinary flat picture frames can be overlaid with vines and fine tracery. Here is an instance where our glue becomes very useful, for we can fasten overlaid work on finished wood, where ordinarily it requires escutcheon pins, but all know it is not pleasant to drive them in light and frail work. Very pretty photograph frames are made by taking pine board and sawing oval out of center and covering pine with velvet. Fasten overlaying on velvet. Designs for overlaying, such as vines or cluster of flowers, a head or any other ornament, can be procured of almost any dealer; but a great many pretty designs for this work can be obtained in such books as Ladies' Book and Harper's Bazaar.

Silhouettes.—Few articles made with scroll saw are more ornamental than Silhouettes. Many designs can be found in books for children. They should be cut from material one-sixteenth of an inch thick, or from veneers. Black and white are favorite colors—ebony and white holly. It requires a tough wood. Finest and best saws should be used. When veneers are used, they should be placed between two pieces of a sixteenth of an inch in thickness each, and fastened firmly. Silhouettes are used in various ways, by overlaying on polished wood or paper placed on a board for a background. If you possess a treadle machine, you can make decorations for your wall, or even make a nice border. Paste a very dark strip around for border, then procure white holly veneer and saw out a variety of patterns—you can saw one-half dozen or more at one time. Glue them on dark border, each equal distances apart. If you wish to saw a perfect likeness of one of the family or a friend, place a piece of paper on wall; with a strong light throw shadow on paper, with pencil trace features and with pantagraph reduce to any size you wish. With little practice you can do excellent work. In this simple way you can decorate an easel you wish to give to a friend with his own likeness, by obtaining it in this way and reducing as stated, placing your picture on wood, saw out and overlay object to be given away.

Sand-Papering.—To sand-paper flat surfaces, always use a block. Two pieces of pine three by four inches, and three-eighths thick; through one of them put a few slender screws, just long enough to come through about one-eighth of an inch, file these points sharp, take piece of sand-paper four inches wide and seven and one-half long, lay one end on screw points, press paper over them, place other block on this, fasten the two together with screws; two sets of blocks are best, one for fine the other for coarse paper; lay work on bench, hold board with one hand and rub with other, give circular motion, move rapidly; begin with No. 1, finish with No. 00. Another good way for small work: lay whole sheet of paper on bench or level board, turn piece to be sand-papered down on it and rub. Small work done nicely in this way.

Fret Sawing in Metals, Shell, Pearl and Ivory.

Brass, gold, silver, shell, ivory and pearl can also be used by Fret Sawyer, many beautiful and useful ornaments being produced by them. It is not generally known, but is a fact, that brass, tin, zinc, and other composition metals, can be cut with the bracket saw almost as easily as wood.

Sawing out thin metals, or thin and brittle substances, the article to be sawn should be placed between two thin pieces of walnut. The design can be placed on wood, as ordinary work, and wood and metal sawed through at same time. By this means very delicate work can be wrought, as wood forms support for thin metal. No matter how fine may be the lines, or how intricate the work, with a steady hand and keen eye the saw will cut hair lines as well as coarse ones.

For metal sawing, only best blades should be used. Nos. 1, 0 and 00 are most desirable. In this manner earrings and various articles of jewelry and ornament may be produced. Two years ago a lad procured one of our Bracket saws. He became an expert at fret-sawing, and is now employed by a manufacturer of jewelry, at $2.75 per day, fret-sawing in silver and gold.

Beautiful ornaments can also be cut from brass, silver, ivory, &c., for inlaying into woods of contrasting colors. Sheet brass, copper, and other metals can be procured of various thicknesses. A silver coin can be hammered quite thin and flat. With the Fret-saw this can be worked up into articles of jewelry, or for inlaying purposes. Silver inlayed into ebony is very beautiful.

Hard, vulcanized India-rubber is manufactured in sheets about two feet square, and is sold by the pound. It can be easily cut, and is very attractive and convenient to use for jewelry, card baskets, &c. Even a handsome clock case can be made from this material. When sawing rubber, frequently place a little oil on blade, to reduce friction.

In putting nice work together, it is of importance to do it well, as good work can be easily spoiled. Hinges can be had at most stores, but they are usually too plain for fine work, and we propose to show how an ornamental hinge can be made by means of

the Fret-Saw. Sheet brass, copper, or other metals, can easily be produced of various thicknesses. The best suited for this purpose will range from $\frac{1}{64}$ to $\frac{1}{8}$ of an inch. Upon a piece of metal, the requisite size, trace, with a fine awl point, the design you have selected. Saw out design, leave flange sufficient to form a turn or socket, where the two parts of hinge are to be united by pin. This flange will at first be straight, but by using a pair of small pliers and working it with light hammer around steel wire fully as large as diameter of pin, it forms a tubular shape. To other half of hinge there will be, of course, two more flanges to be made in same way. Pin should be fitted as true as possible, in order that hinge may open and shut easily. Hinges should be fastened on with small wire pins, holes of proper size being drilled through the metal and wood.

Escutcheons for key-hole ornaments can also be cut from brass and other metals. We illustrate one or two styles. Many other small articles from metals can be cut out with the saw. We illustrate two very useful ones. They are metal loops for suspending

clock cases, picture frames, brackets, &c. For this, sheet brass is used. As nail heads vary in size, the hole by which the loop is passed over the nail should be sufficiently roomy for the largest sized picture nail. Fret-work boxes can be still further ornamented by means of corner braces cut in neat designs from brass. These can be cut in pairs by following the directions already given.

THE TURNING LATHE;

ITS USES AND HOW TO EMPLOY THEM.

THE art of turning, like many others, is simple enough when the fundamental principles are understood, most of which I shall endeavor to make plain by following cuts and explanations accompanying each :

No. 1 shows how to find center. Draw lines from corner to corner; where they cross is centre. On the end you place against tail screw, you must put drop of oil; the other end place against spur center and drive lightly, so as to make spur enter the wood to keep from turning. The tail screw must not be screwed up too tight, as it will make machine run hard.

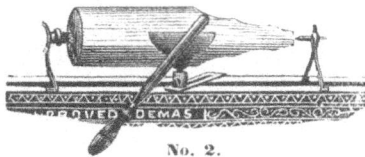

No. 1.

No. 2 shows position of rest and mode of holding gouge to rough off work, the rest must never be below center and always a little above, and the larger the piece you are turning the farther above. To rough off work, take largest gouge, being sure it is sharp. Remember this with all your tools—they must be sharp to do nice work and do it easily.

No. 2.

Observe the slant of gouge ; commence at tail-block and gradually take off a little at a time towards head-block. Should you commence to work at head-block first, you must reverse the slant of gouge. Unless you observe this rule, your tool will catch and run into work. This rule applies to all turning tools, for any and all purposes, and as shown in Nos. 3 and 4, the cutting edges must be kept forward of the hand.

No. 3.

In both of these cuts you will observe especially that the flat chisel is used, and in using this tool care must be taken not to let the upper point of tool get low enough to catch in the piece ; if it does it will spoil the work in spite of you. The lower edge and middle of the tool is used most altogether, unless you wish to cut off a piece—then the long corner is used and the chisel turned over.

In turning round beads, as in Fig. 3, the lower point is used ; in smoothing off straight work, as in Fig. 4, the middle is used ; in cutting off, the long point, with chisel turned over. When the work to be cut off has not stock enough to cut it with the flat chisel, then the cut-off tool must be used. This is the narrow chisel, and to use it, all you do is to place it up against the spot you wish to cut off and force it in perfectly straight and slowly, giving it time to clear its passage ; if forced too rapidly it will leave the end of wood quite rough. These points are the foundation to all turning, and when properly practiced and understood, you can turn anything (with the exception of experience and ingenuity to devise special tools for different fancy turning) ; some of the most valuable I will endeavor to illustrate and explain.

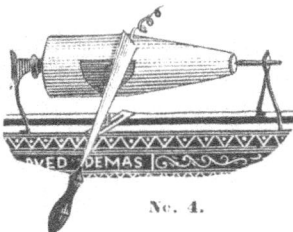

No. 4.

(11)

No. 5.

No. 5 shows you the rosette chuck with a piece screwed on, and rest and tool in proper position for turning it up. For this the tail-block is not used; the chuck is screwed on to head-block and a piece the required size for rosette is screwed on to chuck; then by placing rest in front or on side, you turn any desired design of rosette. Should buttons or small rosettes want to be turned, by having the piece to be turned about two inches long, you can turn one end and then cut it off, and so do until all used up. In this same way, small and short vases, as well as salt cellars, &c., can be turned. To put rosette on to head-block, the nut and centre must be taken out.

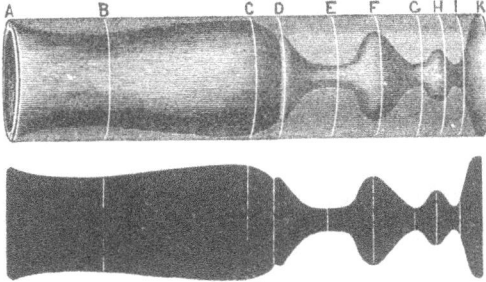

No. 6.

To turn a goblet or any other article to any desired shape, first rough it off and get an ordinary straight surface. Then take a pair of compasses and open them so that each point will touch at the extreme end of pattern, a and k; now press them up against the piece while it is revolving. This will make two rings around piece, which gives full length of article to be turned. Now reduce compasses, and set them at a and b; then hold one point at right hand mark on turned piece, and let other point make a mark to correspond with b on pattern, and continue to do this until all the marks on turned piece correspond with those on pattern. Now take pair of calipers and set them to correspond with some one point, as shown in Fig. 8, and then turn piece down until calipers will slip over it. In this manner you can copy any desired pattern, but after practice you will not require this, for you will be able to do it near enough with the eye.

No. 7.

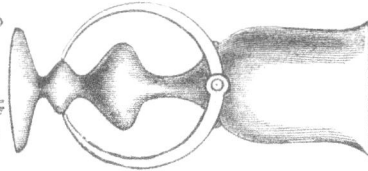

No. 8.

In turning inside of large vases, you must use the tail-screw, as shown in No. 7. To do this, you first take a block the size necessary; then bore a hole in one end the size you want the hole in top of vase and the proper depth; now place this piece in lathe the same as an ordinary piece to be turned, having the hole towards the tail-block; after you have turned the vase in its proper shape outside, place your rest at the lower end and turn out inside, as shown in No. 7.

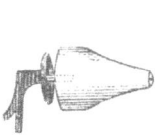

No. 9.

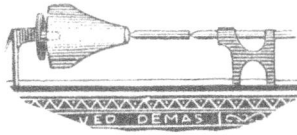

No. 10.

To make a chuck for holding twist drills, take a piece of hard wood about 1½ in. in diameter and two in. long, and bore a hole in one end ⅜ diameter and ⅞ deep; screw it on to head-block and turn it as shown in No. 9, at the same time turning a small center in the end which is free; now take drill such size as you may want to use, place up against this center and bring tail-block up against rear end of drill, as shown in No. 10. Run machine very rapid and press drill into wood slowly so it will enter very centrally; after hole is made, take drill out and wet the hole, and place drill into hole with cutting point out. In this way you can use any sized twist drill and make your own chuck. Each drill must have its own chuck. To turn work after a given pattern, compasses and calipers are used. These can be had at all hardware stores.

DESCRIPTION OF TOOLS

AND INSTRUCTIONS

HOW TO USE THEM.

WOOD CARVING.

ARVING in wood is an elegant and useful art and is easily learned. Of course, elaborate work is not easily done ; but many articles of utility and adornment may be carved by any boy or girl with only ordinary mechanical ability. It is said that whittling is natural to Americans. Carving is only a higher grade of whittling, in which the jack-knife gives place to the chisel and the gouge. It would astonish most people to be shown what may be done with these simple tools. For not only may elegant trifles, such as brackets, book-rests, bread-plates, paper-knives, picture frames, etc., be made by the home carver, but chairs, tables, side-boards, bedsteads, and other domestic articles may be ornamented in this way.

That women may excel in doing carved work, was shown by the work of the Cincinnati Carving Club, exhibited at the Centennial. What has been done by these ladies may be done by other ladies. We are glad to see that carving clubs are forming all over the country, and thereby stimulating hundreds to acquire this elegant and useful art. To aid such as may wish to learn the rudiments of carving, we have prepared the following lessons :

The first lesson of the amateur is to learn the use of the three principal tools—the flat chisel, gouge, and veining tool. (Fig. 1, tools 2, 3, 4.)

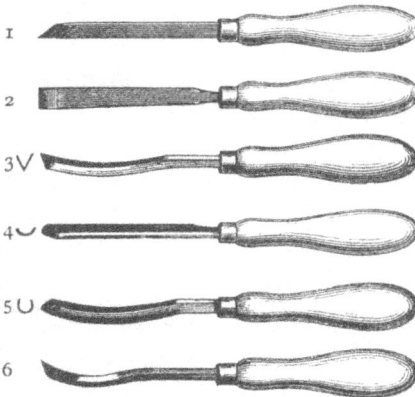

Fig. 1.

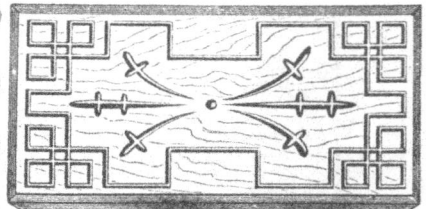

Fig. 2.

Fig. 3.

For this lesson, take a block of white wood, six inches long by two or three broad and one and a half thick. Secure it firmly to a bench ; then, with the flat chisel, carve the beveled edges and make the mitres perfect. Now take the veining tool, No. 3, which cuts a V shaped groove, and carve out the design, as seen in the cut (Fig. 2).

In using this tool, it must be held in the right hand and in a slanting direction. The left hand should be hollowed and placed on the tool, the wrist and tips of the fingers resting upon the work. This steadies the right hand and prevents the tool from slipping forward. Now use the gouge, and carve out the circular depression which may be seen in the design. Practice on this lesson until you can carve the design accurately.

Fig. 3 shows a slip of Arkansas stone for sharpening the veining tool, No. 3, and gouges.

After you have mastered these tools, so as to use them with ease, you may attend to the direction for carving a wall pocket (Fig. 4).

Fig. 4.

Fig. 5.

Fig. 6.

The wood to be carved should be black walnut, 12 x 14, and one-half of an inch thick. It must be well seasoned, straight grained, and free from knots.

First, give the outline shape of the design, as seen in the cut (Fig. 4), which may be done with an ordinary fret saw. Then sketch the pattern upon the wood. A knowledge of drawing will greatly aid the carver, yet in many cases the drawing can be easily made by means of impression paper. Place the wood upon a table or bench, and secure it firmly by means of a clamp or screws. With the parting tool, No. 3, cut the V shaped groove for the outline of the design.

Having finished the grooving, some of the carving punches (Fig. 5 shows impressions made with five different styles) are next to be used. First use the punch *c* for making the three circular impressions in each corner. Then with punch *b* go over the design as in Fig. 4. The cross-shaped punch, *d*, can be used to ornament the outer rim. Very handsome ornamental work can be done with only the punch *b*, and for this design it can be used to good effect without the aid of the other punches.

In using the punch, hold it perpendicularly in the left hand, and with a mallet give it a sharp, quick blow. Furniture ornamented in this style looks remarkably well, and the work is easily executed.

RELIEF CARVING.

Rim of Bread Platter, ornamented with "Relief Carving."

Our next lesson will be to carve the end of a book-rack with the grape-leaf pattern in relief (Fig. 6). The wood, either walnut, oak, or mahogany, should be half an inch thick and six inches wide ; but of course the proportions and sizes may vary to suit the wishes of the carver. Having sketched the design upon the wood, the next thing to be done is to "stab out the work." This is accomplished by holding the chisel, No. 2, upright on the line and pressing it downward to the depth of about one-sixteenth of an inch. It is better not to cut exactly in the line, but keep just outside.

When the "stabbing" has been done, hold the chisel slantingly, and cut towards the pattern, thereby removing the wood near it, and leaving it quite free. Then with tools No. 4 and No. 6, clear away the "dead" wood in the intermediate space, leaving the design standing up in relief. You must notice carefully the grain of the wood. It the grain runs downward, turn the wood around and work the reverse way, or side-ways. You will readily apprehend that the whole of the branches and leaves which form the pattern should not lie in one dead level. The thickest portion of the stems, leaves and grapes should be the highest ; the rest of the design must be harmoniously lowered. At this stage you must remove all those parts which the design indicates are to lie very low. You must be careful as you go on to " stab out " the outline again and again, so that you may keep accurately to the copy. Next, you must " stab out " the whole design again, and deepen the ground-work until it is about a quarter of an inch lower than the upper surface. You may then carve the leaves, stalks and grapes, copying nature as far as possible. In doing this, you will have occasion to exercise your own judgment and taste.

CARVED FRET WORK.

Another popular style of carving is "carved fret work." For this lesson we will take a carved paper-knife (Fig. 7).

Fig. 7.

Take a good piece of black walnut, ten inches long, one and one-half inches wide, and one-quarter of an inch thick. On this trace the design as seen in Fig. 7. With your fret saw cut out the pattern, as is done in ordinary fret sawing. Now, with your tools carve the design to imitate as closely as possible the natural leaf, flower and stalk. This being done, with sand paper and a knife, bevel the edges and shape the handle.

Fig. 8 shows a specimen of fret carving. The dark portion of the bracket shows it simply sawed out with the bracket saw. The other side has been carved.

Fig. 8.

Design No. 9 is a carved moulding, in which the leaf and stem are to be left of the original height of the wood, the other parts being cut away ; and the veins are cut with the parting tool, No. 3. It should then be punched with one or the other of the punches.

Fig. 9.

It is necessary that all tools used in wood carving should have sharp, keen edges. Do not try to carve with dull tools.

PRIZE HOLLY SCROLL SAW.

NEVER before was there a machine which, in so short a time, gained such a world-wide reputation ; its name is as a passport to every scroll sawyer. Now that we have changed its appearance, we distinctly wish to say that none of the good qualities have been lost ; but, to the contrary, the change has been made for the benefit of great improvements. The rapid progress made for the benefit of amateurs in the past five years has produced the following results.

1st—That there are three classes to supply.

2d—That all want machines similar in construction, but vary materially, to differ in prices.

The very young and those of small means for the first. The second, the medium in years and that which pertains to riches.

3d—Those who have become experts with the first or second machine, and now are able, from the earnings of the same, to invest in that which is not only better but will execute larger and more elaborate work. For prices, see page 31.

The sale of these machines varies according to the price; the lowest having by far the greatest sales. There are several reasons for this: First, its price. Second, the large number of youths. Third, and best, is that it answers most every requirement. It is an easy matter to make a machine and have it give perfect satisfaction, so long as price is not stipulated; but to make a machine at the price the PRIZE HOLLY is offered and give it all requirements and have them give perfect satisfaction, requires some thought and careful manipulation. But my patrons will find that the greatest success has been attained in this respect with the above machine. Its strength, durability and ease of operating far excels anything ever before attempted. Its simplicity of construction has never been equalled. It is adjustable in every conceivable shape, so that the most inexperienced can set it up and make it work perfectly.

It is especially constructed so that it can be used for a practical Lathe. This latter feature is the cause of its change in appearance, which adds not only to its value, but has also given us a much better saw than when in its former state. In the above we give an illustration of the machine when in use as a Lathe only. Now, to change this back to a Saw requires but the moving of the rest to the right and placing Saw attachment on the Lathe bed and fasten with one screw—the work of but a few seconds.

The beauty of this machine is, you can buy the Saw only, or the Lathe only; or you can buy both combined. The latter is always the most practical, if you ever expect to have both, as it saves freight. But if either are bought separate, at any time thereafter you can buy the balance at no extra expense except freight. I wish especially to call the attention of parents to the fact that this machine is especially adapted to the very young beginner, and if he can not learn on this there is not a machine built that he can; and do not for a moment think because it is cheap it will not answer the purpose just as well, for I guarantee you it will do just as fine work, and do it just as well and just as easy, as any other machine built; its only difference is in plainer construction and a trifle smaller capacity. Its dimensions and capacity for work are as follows: height of machine, 30 inches; full width, 18 inches; diameter of balance wheel, 12 inches; weight, 7 pounds. Scroll Saw will cut 1½ inches thick and swing 20 inches in the clear; stroke, 1½ inch; with it you can use the very finest or coarsest blade. The Lathe will turn 10 inches long and 4 inches in diameter. Small chucks, for holding drills, &c., can be fitted to spindle if desired. The Lathe-bed ways are ground and polished, which gives a smooth and level surface for head and tail block to travel on. The tilting table, for inlaid work on Scroll Saw, is also ground and polished. We feel safe in saying that the Prize Holly will more than double its former popularity. Full weight of machine, 30 pounds. For prices, see page 31.

Prize Demas Lathe and Scroll Saw,

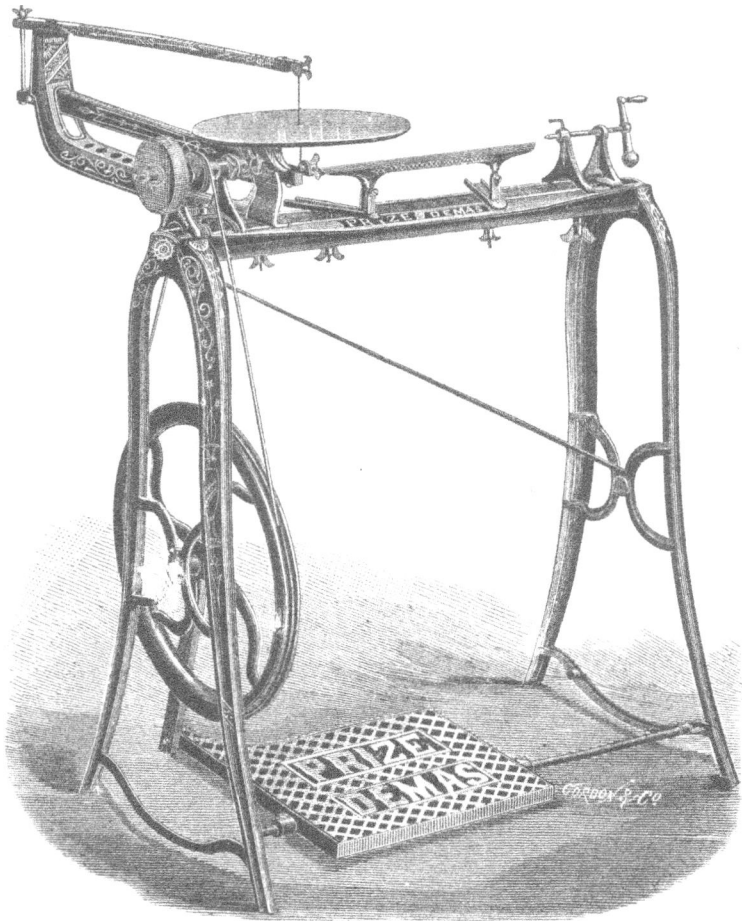

BY FAR THE MOST POWERFUL AND PRACTICAL, EXECUTING THE

LARGEST AND GREATEST VARIETY OF WORK

OF ANY IN THE MARKET.

For description see opposite page.

Description of Prize Demas.

THIS machine, as its name indicates, will be a prize to any one fortunate enough to possess one of them. Its principles are such as practical experience have proven to excel in every particular. Five years of constant devotion to the public's wants has produced this result ; in it we have overcome every defective point which may have been brought up by some of the many thousands who, in the past five years have used a lathe and saw of some description, and we can now truthfully say that in this machine you get everything that the heart may crave for, and put up in such a manner as will give you perfect satisfaction. It is as much a necessity in every household as the sewing machine, and its benefits are much greater. Every man can be proud of it. Every machine or part thereto is thoroughly inspected before leaving the factory. Still, as heretofore, we consider it a favor for any party to notify us at once, should they get a machine which is not perfectly satisfactory in every respect, as we have always made it a point to please all and wish to continue to do the same.

The above cut is an exact representation of the true proportions of this machine, which is exceedingly symmetrical and tasty in appearance. While we have been sparing of the iron, it has been properly and practically distributed, producing the greatest possible strength.

Dimensions and capacity of machine are as follows :—Height from floor to top of lathe bed, $27\frac{1}{2}$ in.; to centers, 30 in.; to top of saw table, 32 in.; length of lathe bed, $24\frac{1}{2}$in.; it will turn a piece 16 in. long and 5 in. in diameter; diameter of balance wheel, 14 in.; weight, 11 lbs.; stroke of crank, 4 in,; size of lathe spindle, $\frac{7}{16}$; short rest, 4 in. long; long rest, 12 in. long; stroke of scroll saw, $1\frac{3}{4}$ in.; it will cut $1\frac{1}{2}$ in. thick if necessary, but 1 in. practically, and swing 20 in. in the clear; it has tilting table, which is ground and polished ; the ways to lathe bed are also ground and polished—in fact, it is so finished that it will do just as fine work and just as satisfactorily as any $50.00 lathe and saw. Chucks for holding drills, &c., can be attached. Weight, 50 pounds. For prices, see page 31.

Illustrations of Wood Type

MADE ON PRIZE DEMAS AND HOLLY, FURNISHED ON APPLICATION.

The utility of these machines for executing the most intricate patterns in fret-sawing is attested in the immense sale they have had, over 40,000 being sold by one firm. This and the fact that I had, for the past years, made hundreds of wood type for my own printing purposes, induced me, about 1 year ago, to offer them, with proper material, to printing establishments, that they might make many of their own type. Since then I have sold hundreds of them, and in every instance they have given perfect satisfaction. The very close grain of the Holly wood makes it equal to an electrotype, for printing purposes. I have type from which over 300,000 impressions have been taken, and they do as good work to-day as when first made. For those who do not understand how it is done, I would say that the Holly I keep in several thicknesses, $\frac{1}{16}$ and $\frac{1}{8}$ being used for printing. Now the design you wish is placed upon the Holly in reversed position, the same as you set it up. You can saw either 1 or $\frac{1}{2}$ doz. at a time. After it is sawed you glue it upon pine that is of proper thickness to make it type high. Many object at first, thinking that as they wash the type it will soften the glue, and the Holly come off ; but this we understood, and we furnish you a glue that will not do this. We would further state that any boy can learn in a short time to make type. It will more than save its cost, in a few months, for mortising alone. Either Holly or Demas can be used, but Demas is preferable, as it is heavier and will do its work with greater rapidity. For prices of machines and printers' material, see page 32.

(19)

DEMAS LATHE, NO. 4.

A Practical Lathe for Metal and Wood.

Designed expressly for machinists, carpenters, cabinet makers, dentists, pattern makers, and those who make honey boxes. For description, see opposite page. For prices, see page 31.

Our Large Lathe and Scroll Saw,

DEMAS No. 4 AND DEMAS No. 5,

With their attachments, have great power and mechanical capabilities, and being constructed on a large scale, with regard to correct mechanical principles, the prices are so low that an increasing trade keeps pace with our advertisement of their respective merits.

When mechanics, using other machinery to power; mechanics away from power and the amateur in his home workshop, learn that such large, well finished and substantial machines, for strictly manufacturing purposes, *can be bought at less figures than nickel plated Amateur machines*, of no finer finish and accuracy, orders are mailed us so that our manufacture of these implements has become in two years from an experiment an enlarging proportion of our trade.

Description of Demas No. 4 Lathe.

Built of Iron and Steel; Treadles of Ash.

Length of Lathe Bed, 42 inches ; Ways are planed. Height from floor, 34 inches ; to Centers, $39\frac{1}{2}$ inches ; Centers from Lathe Ways, 5 inches; greatest distance between Centers, 30 inches.

Head and Tail Blocks are milled to fit Lathe Ways.

Spindle for Head Block is of $\frac{3}{4}$ steel.

Boxes are adjustable, so as to take up any lost motion that may occur from continued wearing.

Centres are fitted with taper shanks.

Set Screw for Tail Block spindle and small *Balance Wheel for Head Block, not shown in cut.*

Chucks of all descriptions can be applied.

The Cone Pulley on Head Block is nicely turned up and has two lifts to correspond with two grooves on large Balance Wheel.

Shaft for wheel, $\frac{6}{8}$ steel.

Five-sixteenths ($\frac{5}{16}$) brace, full length of machine, for firmness.

Adjustable Lathe Rest and Tail Block have hand wheels.

Five-sixteenths ($\frac{5}{16}$) belt 86 inches long, and wire fastening with each Saw or Lathe.

Diameter of heavy Balance Wheel, 21 inches ; full weight of Lathe, 130 pounds.

CAPACITY.

Is sufficiently large and powerful to turn table legs, pillars, posts, chair legs, spindles ; also firm and accurate for purposes of finer work, while with our "Universal Slide Rest," screws and articles of metal may be turned.

With the Scroll Saw attachment (see No. 5, page 23), which also fits this machine (see price list), building brackets, wagon felloes, fancy brackets, &c., &c., can be made with very little labor, time and expense. Sawing such by hand is tiresome and slow compared with the easy and swift execution by this very cheap attachment.

Also, the useful attachments described on pages 24-5 are every day requirements to the mechanic and amateur, and instead of each being separately constructed with frames, gearing and power, the parts that directly perform the various kinds of work are made as attachments, and adjusted to either Nos. 4 or 5 Demas, thereby combining six or more machines in one.

This machine is a companion in time of need to mechanics away from power, and farmers are learning its value. Can also be operated by steam.

Price of Demas No. 4, Lathe only, page 31.

UNIVERSAL SLIDE REST.

THIS is an iron turning attachment for Demas No. 4, and is intended for turning up pulleys, small pieces of castings, and short lengths of shafting ; it is nicely fitted up, and will execute work in the most satisfactory manner. Capacity of cross feed, 2 inches ; horizontal feed, 5½ inches. It can be so attached to the Lathe-bed as to face off any sized pulley that can be chucked in the Lathe. It is the largest and most perfect slide rest ever made, for the money, and can be attached to any Lathe. Weight of slide, 15 pounds. For prices, see page 31.

THE SHIPMAN ENGINES, Nos. 1 & 2.

One Insertion of our Advertisement in an Eastern Paper Results in a Deluge of Letters and Postals containing inquiries that are answered on this Page.

THE SHIPMAN ENGINE, No. 1.

CAPACITY.—Any foot-power machine or contrivance, or as many machines that can be propelled by foot-power, however ingenious or labor-saving the gearing may be. It will run a machine quietly, steadily, and with great speed ten hours, for a few cents, that would tire a man in four minutes to equal, if he attempted the same momentum. The uses for which this Engine is applicable are enumerated on third page of Engine Circular ; of course there are many others, among which are a long list of machines and geared appliances operated by hand-power. Therefore, *any implement that can be AT ALL worked by foot-power, or any thing operated by hand crank, which a man can turn two hours without exhaustion, my No. 1 Engine will drive.* We hope that by closely scanning the above, our correspondents will be enabled to determine the adaptability of the No. 1 Engine to their particular work. We desire to add that it has been run in the factory one year, with results that justify the assertion that they are lastingly durable, and we certainly could not afford the costly experiment of making them for purposes in which they would fail.

SHIPMAN ENGINE, No. 2.
Price, including Oil Tank, ...$75.00.

Has been perfected since issue of Engine Circular. The principles that have made No. 1 Engine a complete success are implied in this : appliances for regulating water, flame, steam, speed, pressure, &c., &c., with mechanical accuracy, and are fully up to the essential parts to any engine. With a few additional principles and more elaborate construction than the smaller Engine, No. 2 Shipman Engine, has greater capacity, which we advise our correspondents to estimate by what a one-horse power will do, as the No. 2 really exceeds this. It will be found serviceable in cases where the No. 1 is inadequate. Where the No. 1 will exceed foot-power somewhat, the No. 2 runs circular saws, moulding knives, larger boats, heavier lathes, a greater number of manufacturing sewing machines and printing presses. Also, pumping water for irrigation ; also, fans for forges with a great blast.

A Board of Underwriters has, on a rigid examination, decided these Engines to be " no extra risk." Farmers need have no fears to take them into barns and granaries, as the flame is inclosed, but will warm a small room.

Weight of No. 2 Engine, 125 lbs ; dimensions are 2 feet high, and 24 x 18.

Attachments for Sewing Machines, enabling the operator to start or stop by pressure of knee..................... price, $5.00.

1 1-4 inch Belt,....................   "   12 per foot.
1 inch Belt.......................   "   10 "

No. 1 Engine, the smallest ; and No. 2 the largest Engine made at present.

BUZZ SAW ATTACHMENT
For Nos. 4 and 5 Demas.

In ordering above, please state what machine it is needed for, No. 4 Demas, No. 5 Demas, or New and Improved No. 5 Demas Scroll Saw.

Dovetailing Attachment for Demas No. 4.

DOVETAILING ATTACHMENT.

The cut illustrates the utility of this machine full as well, if not better, than words can explain it. Its merits can readily be appreciated by those who wish to put together in a very rapid, practical, neat, and substantial manner, small packing boxes, honey boxes, drawers of every description. The table is adjustable for cutting any depth. Its principles are such that any width board can be dovetailed. Four saws, $\frac{1}{8}$ thick, accompany each machine, but more can be used if desired, but is not necessary. For prices, see page 31.

Buzz-Saw Attachment for Demas Nos. 4 & 5.

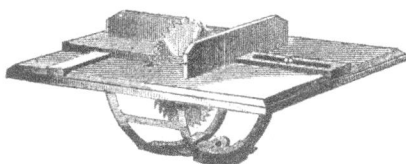

BUZZ SAW ATTACHMENT.

This attachment, as cut indicates, has both a cross-cut slide and a ripping gauge. The table is 15 x 18 inches, and has a 6-inch saw. One saw accompanies each machine, either cross-cut or rip-saw; will be sent according to order; but when the order does not designate which, we always send rip, as it is most practical for general work when one is used only. The table is adjustable up and down, so that when the saw is made to wabble, grooves of any desired width or depth can be cut. For prices, see page 31.

Moulding Attachment for Demas Nos. 4 & 5.

MOULDING ATTACHMENT.

The value of this machine is too well known to all workers in wood to need much description; its dimensions and capacity of work are as follows: the table is 10 inches in diameter, turned up true and then polished, which gives a very nice and even surface for wood to move on. The spindle is made from $\frac{3}{4}$ steel, and nicely fitted up; and $\frac{3}{4}$ knives can be used, which, by foot power, is as much as can very easily be done; speed can be varied from 2,500 to 3,500 revolutions. We do not furnish any knives with this machine, for the reason that we could not tell the shape required. Knives are furnished extra, and a drawing showing the shape of knives wanted must be sent when ordering. For prices, see page 31.

Grinding and Polishing Attachment,

FOR DEMAS Nos. 4 and 5.

GRINDING & POLISHING ATTACHMENT.

The cut represents this attachment as especially adapted for jewelers' and dentists' use. Emery wheels can be applied between nut and collars at one end, and the taper screw is for brushes and buffers of every description. It is nicely fitted up and can be run at a high rate of speed. We also fit up the same head with heavier spindle and put nut and collars on both ends, especially adapted for emery wheels for grinding purposes. In this shape it is invaluable for small shops of every description, and farmers will find it the best Reaper-knife grinder ever made. For prices, see page 31.

Buffer and Polishing Attachment,

FOR PRIZE DEMAS AND HOLLY.

TAPER SCREW.

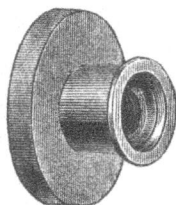

EMERY CHUCK.

This is a brass taper screw, and is screwed on and off spindle in same place as spur center.

The other cut represents a brass chuck for holding small emery wheels. Wheels are fastened to it by shellac. This is attached to spindle same as taper screw. The beach almond or any other chucks can be applied in same manner, and with these attachments the Prize Demas and Holly make the most perfect and complete jewelers' and dentists' lathes that have ever been placed upon the market. Many are already using them with great satisfaction, and with these improvements we feel sure hundreds will avail themselves of this opportunity of getting a most complete machine for very little money.

These can be sent by mail. For prices see page 31.

Prize Demas & Holly Scroll Saw Attachment.

PRIZE DEMAS AND HOLLY ATTACHMENT.

This cut shows this attachment as it appears when detached from machine. The principles of this saw are the most practical and perfect of any ever built and are so simple that any one can learn to use it, and so adjust it as to make it work perfect. For prices, see page 31.

Shipman's Shuteing Board, with Amateur Plane.

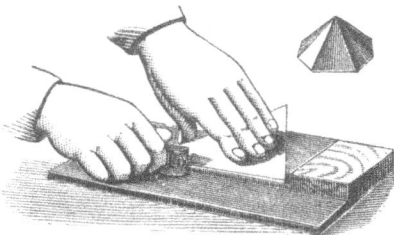

When making Octagon Tops.

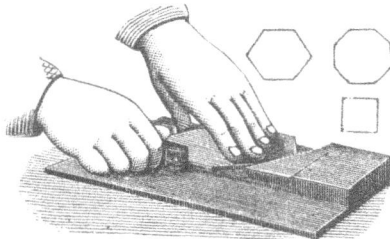

When making 6 and 8 Sq. Bottoms.

Iron Amateur Plane.   1¼ in. wide.   3½ in. long.

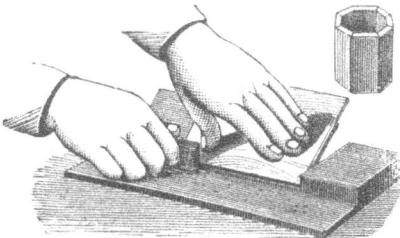

When Beveling 6 and 8 Square.

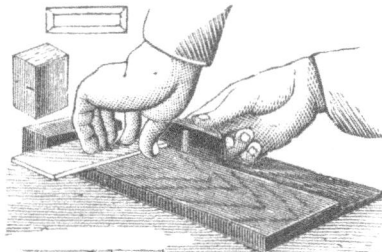

When making Square Joints.

The above cut represents the *Board*, with its Attachments, placed in position for doing the different kinds of work, full instructions with each board, all put up in a neat box, and sent to any address at the prices given on page 32.

This not only essential for all Amateurs, but is indispensable to every Carpenter, Cabinet Maker, Pattern Maker, and in fact, Wood Workers of every description.

Fret Sawyers and Lovers of Fancy Work cannot do without it.

SHIPMAN'S CARVING TOOLS

Are made of the Best Quality of Steel.

SIX CARVING TOOLS.

ONE PUNCH.

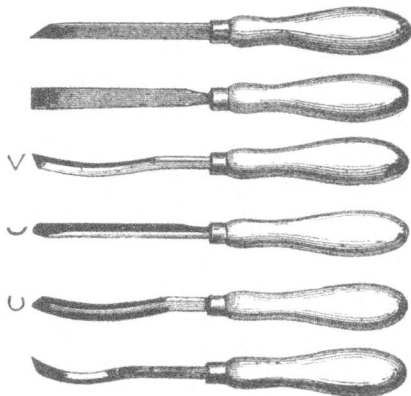

MANUAL OF CARVING.

With this outfit, all can learn the fascinating art.

ALSO A COMPLETE

For prices, see page 31.

GRIFFIN PATENT SAW BLADE.

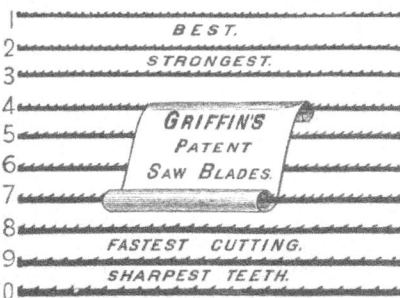

1
2   BEST.
3   STRONGEST.
4
5   GRIFFIN'S
6   PATENT
7   SAW BLADES.
8
9   FASTEST CUTTING.
10   SHARPEST TEETH.

The following is worth reading. It is from no boy, and is not a purchased commendation. We have submitted these blades to the severest test, and cannot commend them too highly.

PLEASE READ.

GENTLEMEN: I have thoroughly tested the GRIFFIN PATENT SAW-BLADES, on various kinds of wood and metal, and I am convinced that they are *stronger* and *more desirable* in every particular than the *French* or *German* blades.

It is well known that the imported saws have a BURR on one side of the blade, which causes them to run off the line, unless care is taken. The Griffin Blades run perfectly true, as each tooth is set. No other blade has this very desirable improvement. I have cut soft sheet-brass, one-fourth of an inch thick, with the Griffin Blades, with ease. In fact, with the Griffin Blades, and the Prize Holly Saw, I can make fine patterns of wood or metal for casting better and quicker than with any other tool. For *strength*, *durability* and *rapid cutting* I consider the Griffin Blades far superior to any other I have seen.

Yours truly,

C. H. THURSTON,
Treasurer Thurston Knob Screw Co.

WE keep no other blades in stock, for the reason that when once these are used we cannot sell any other It is superior to all others on account of the accuracy of the cut; it will follow a line with less trouble than any other, for the reason that it has set on both sides, and does not have that rough burr on one side, as found on all imported blades; its looks deceive it, and needs to be tried to be appreciated; it is much *stronger*, from the fact that the body has more *material* in it, its teeth are much *sharper*, and you do not have to *press* so *hard*, which causes so much more *friction*. These saws when *ordinarily* dull are *sharper* than most of the old style blades; another great advantage is that they *are all alike*, which never was known in other blades. Try them and see. For prices, see page 32.

BRASS HINGES.

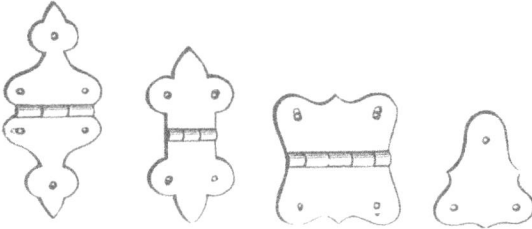

This cut shows some new designs in Hinges, made expresssly for fret work. These Hinges, with the fine French screws we now can furnish, fill a want long felt by the lovers of this work.

For prices of Hinges and Screws, see page 31.

BRASS HINGES.

IMPROVED BRACKET-SAW DRILL.

This Drill is fitted up in the very best manner. Its parts are all made of hardened steel. It has a very fine chuck, and fret workers will find this a very useful article, for many holes can be made with it that you can not get at with power drill. For prices, see page 31.

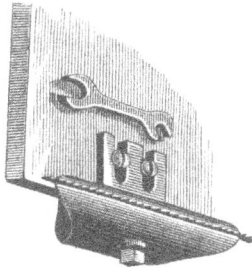

DRILL.   SAW FILER.   COARSE SAW BLADES.

GIANT SAW FILER.

This little device will save its cost in one week, and those who have tried it would not part with it for five times its cost. In doing heavy work it is necessary to use a saw with moderate coarse teeth, and it is necessary they should be sharp to do their work with ease and give satisfaction. This machine will do this perfectly. Prices on page 31.

LONG AND COARSE SAW BLADES.

These blades are especially adapted for Nos. 4 and 5 machines, but can be used on Prize Demas and Holly if so desired; for heavy work no others should be used. Prices on page 31.

AMATEUR CHUCK, FOR LATHES.

This chuck is especially adapted for amateurs' use : first, because it is the cheapest ; second, because it is the strongest and simplest, and will not get out of order as easy as most all others ; third, and best, is because it will do the largest variety of work, this being the most essential point to all amateurs. It will hold any sized drill up to ½ inch and will chuck a pulley or other piece of work up to 2½ inches in diameter ; pulleys even larger than this can be held for boring out or turning off. This chuck can be fitted to any of my Lathes and much larger chucks can be fitted to No. 4 Lathes We can also fit any other chucks that parties may desire. Prices on page 31.

This cut represents a window garden scene—the decorative work all having been done with the New and Improved Demas Lathe and Saw. This embraces a variety of work which never before has been executed with an amateur lathe and saw, and

is certainly a credit to any machine that can do it. It would cost more than twice the price of the machine to get this work made, but with this machine the cost is but a trifle. Any of the patterns, separate, for sale.

WINDOW gardening has become too popular to need any comments. Many devices have been introduced to shield the flower pots and increase the beauty of the scene. The above cut represents one of those, which, judging from its sales in the past year, places it far in advance of anything else ever produced. The beauty of its effect is very nicely displayed in this cut and it is so simple that any scroll sawyer can make it. For prices, see page 32.

THE LORD'S PRAYER

IS a pattern that every fret sawyer should aim to have among his many fine specimens of work; it is one that should be in the house of every one that has a Bracket Saw. The trouble heretofore has been to get one that was very handsome and at the same time easy to saw. This I have accomplished in this pattern. It can be sawed from one piece or from small pieces. In ordering, mention Lord's Prayer, No. 2. For prices, see page 32.

Full size Working Pattern, 15 × 18.

PRICE LIST

Mention 1881 Catalogue, when Ordering Goods from this List.

IN ORDERING, CARE MUST BE TAKEN TO STATE JUST WHAT YOU WANT AND PRICE OF EACH ARTICLE.

BE SURE AND READ THE FOLLOWING.

You will notice there are Nos. 1, 2 and 3 Prize Holly, and Nos. 1 and 2 Prize Demas. Now there is no difference in the machine itself ; what makes the difference in price is the extra articles that go with them, as enumerated. The reason we do this is so that those living out in the country, away from where woods, patterns, &c., can be bought, can get these things at the same time they do their machine ; for it costs no more to get a No. 3 machine than a No. 1, that is by freight, for they charge for 100 pounds anyway, even if it is only 25 pounds. You will notice we give prices by mail for woods for fret-sawing. Many who live at great distances will find this very handy, as they can get small lots so much cheaper than by express, but only refers to small quantities. Postage stamps will be taken for small amounts, but always send one-cent ones where it is possible. Stamps that are stuck together, or fast to the letter, are of no use, and will be returned.

A large sheet of miniature patterns will be sent free to any who may wish to buy patterns. Those wishing such will be kind enough to state miniature sheet. I will deem it a special favor if parties, after using my machine for one or two weeks, will write me how they get along and what they have accomplished with it.

PRIZE HOLLY.

Prize Holly No. 1 consists of Scroll Saw, with Polished Tilting Table, Emery Wheel and Power Drill...............Price $3.00; with Lathe and 3 Turning Tools, $5.00
Prize Holly No. 2, same as No. 1, with 12 Saw Blades, 3 Drill Points and 30 Designs extra,..... 3.50
   With Lathe and 3 Turning Chisels 5.50
Prize Holly No. 3, same as No. 2, with Book of 87 Patterns and 15 ft. assorted Wood extra... . 4.50
   With Lathe and 3 Turning Tools..... 6.50
Prize Holly Turning Lathe, with 3 Turning Tools,..... 3.50
   Scroll Saw Attachment, $1 50; Lathe Attachment, $2.00; Buzz Saw Attachment, with one 3¼ in. Saw, $1.00; with Dovetailing Attachment and 1 Saw, $1.50.See bottom of page.
   ☞ A most complete Illustrated Manual on Scroll Sawing, Turning and Carving, accompanies each.

. . PRIZE DEMAS.

Prize Demas No. 1 consists of Scroll Saw, with Polished Tilting Table, Emery Wheel, Power Drill, 12 Saw Blades, 3 Drill Points and 30 patterns, $5.00; with Lathe and 5 Turn'g Tools, 8.00
Prize Demas No. 2, same as No. 1, with Book of 87 Patterns and 15 ft. assorted Wood extra... 6.00
   With Lathe and 6 Turning Tools,..... 9.00
Prize Demas Turning Lathe, with 5 Turning Tools..... 6.00
   Scroll Saw Attachment, $2.00; Lathe Attachment, $3.00.
Prize Demas Buzz Saw Attachment, with one 3¼ in. Saw, $1.50; with Dovetailing Attachment, and one Saw 2.00
   ☞ For Cabinet Prize Demas, add $10.00 to above prices.
   ☞ A most complete Illustrated Manual on Scroll Sawing, Turning and Carving, accompanies each.

DEMAS LATHE Nos. 4 and 5, and SCROLL SAW No. 5.

Demas Turning Lathe No. 4, no extra tools; price.... ..... 15.00
   Tools for No. 4 Lathe; price, each..... .25
Demas Scroll Saw No. 5 (no extras with this machine); price..... 12.00
Dovetailing Attachment, with four Saws (this only fits on No. 4 Lathe); price..... 10.00
Universal Slide Rest, for No. 4 Tools being extra, 50 cents each..... 15.00
Buzz Saw Attachment, with one 6-inch Saw (will fit either No. 4 or 5 machine); price.......... 5.00
Scroll Saw Attachment, no Saws (will fit either No. 4 or 5 machine); price 5.00
Moulding Attachment, no knives (will fit either No. 4 or 5 machine); price..... 10.00
   Moulding Attachment Knives, per set..... 1.00
Buffing and Polishing Attachment (will fit either No. 4 or 5 machine) 3.00

JEWELERS' AND DENTISTS' ATTACHMENT FOR PRIZE DEMAS AND HOLLY.

Taper Screw for Buffing Attachment to Prize Holly and Demas, ... .50
Chuck for Emery Wheel to Prize Holly and Demas, ... .50

SHUTEING BOARDS.

Whitewood Shuting Board, with Plane, 75c. ; same, Black Walnut. $1.00; by mail, 25c. extra.

Small Plane, separate, 25c.; by mail, 5c. extra. | Giant Saw Filer, 25c.; by mail, 15c. extra.
Carving Tools, per set, ... $1.00 | Impr'd Bracket Saw Drill, $1.00; by mail, $1.10

SHIPMAN'S MANUAL AND PATTERNS.

This is a book of 87 full-sized patterns; it also contains full instructions in Fret Sawing, giving description of all the woods and their uses; also has a treatise on gluing, shellacing, polishing, sand-papering, &c. In fact, as a manual, you get all that is in books that cost $1.50. Price of Manual and Patterns, 50 cents.

GIANT GLUE.

This is a Glue prepared expressly for Fret Sawyers, and there is no other glue ever placed upon the market that gives as good satisfaction. It is white, always ready to be used, and will glue any thing in the shape of wood, glass or crockery ware. It will also glue silhouettes on varnished work. Those who have once used it will use no other. Cannot be sent by mail. Price per bottle, 25 cents.

(31)

In ordering Buzz and Dovetailing Attachment separately, state name of your machine, and if it has a Lathe attachment.

PRICE LIST.

Mention 1881 Catalogue, when Ordering Goods from this List.

IN ORDERING, CARE MUST BE TAKEN TO STATE JUST WHAT YOU WANT AND PRICE OF EACH ARTICLE.

CLOCK WORKS.

Clock Works— 8 day Pendulum, Nickel Plated Dials, $2.00; Postage, ...30c. extra.
" " 24 hour " " " " 1.50; " ....25c. "
" " 30 hour Lever,............................1.60; " ....25c. "
" " 30 " " Striking, ...2.00; " ....25c. "

SAW BLADES.

Griffin's Patent Saw Blades—Nos. 0, 1, 2, 3, 4, 5, 6, 7,....15c. per doz.; Nos. 8, 9, 10, 20c. per doz.
" " " " Same Nos.,..................$1.25 " gross; Same Nos. $1.50 " gross.
Saw Blades for Demas, No. 4, 7¼ in. long, A and B, 10c. each; 75c. per doz.; D, E and F, 15c. each; $1.50 per doz.

IMPRESSION PAPER, FOR TRANSFERRING PATTERNS ON WOOD.

Impression Paper, black on one side, 8½ x 11, 5c.; 11 x 17,.........10c.; 17 x 22,.............15c.

SMALL BENCH VISES.

No. 1 has 2-inch jaw, open 2 inches, and has Swivel Attachment; price.............. 1.25
No. 2 has 1½-inch jaw, open 1½ inch; price.............. .75

AMATEUR DRILL AND LEVER CHUCKS.

Amateur Drill Chuck, will hold ⅛ inch drill and under, fitted to either machine.............. 5.00
Trump Drill Chuck, will hold ¼ inch and under, fitted to either machine.............. 2.50
Drills for Demas and Holly machines, from 1-64 to ⅛ inch, 5c. each, per doz.............. .40
Perfect Dust Blower, to fit any machine.............. .25

BRASS HINGES AND SCREWS.

Three Assorted pairs of small Hinges, per doz.............. 40c.
" Screws, per doz.............. 25c.
" Escutcheon Pins, per hundred.............. 25c.

OIL STONES.

Best quality Knife Blade Slips, for carving tools,.............. 50c.
" Round edge " " turning tools.............. 65c.

FILES.

Round, half-round and flat, for wood, and 4 inches long.............. 15c.
3-Square, for filing saws.............. 20c.

BUZZ SAWS FOR PRIZE DEMAS AND HOLLY AND Nos. 4 and 5.

For Prize Demas and Holly, 3½ inches in diameter, each.............. 25c.
" " " Dovetailing, each.............. 25c.
For Demas Nos. 4 and 5, 6 inches in diameter.............. 1.00
" " Dovetailing.............. 50c.

SAW CLAMPS, BELTS AND BELT HOOKS.

Saw Clamps, for either machine, per pair.............. 25c.
Belt Hooks, " " " doz.............. 10c.
Belts for Prize Holly and Demas, each.............. 20c.
" Demas Nos. 4 and 5, each.............. 50c.

PRICE LIST OF WOODS FOR FRET SAWING, PER SQ. FOOT.

			By Mail.				By Mail.
White Holly,	1/16 thick,	10 cents.	15c.	Red Cedar,	1/8 thick,	10 cents.	16c.
" "	1/8 "	10 "	17c.	" "	3/16 "	12 "	21c.
" "	3/16 "	12 "	23c.	Bird's-eye Maple,	1/8 "	15 "	23c.
Black Walnut,	1/8 "	6 "	12c.	" "	3/16 "	20 "	33c.
" "	3/16 "	7 "	18c.	Cocobolo,	1/8 "	20 "	33c.
" "	1/4 "	8 "	20c.	Rose Wood,	1/8 "	20 "	29c.
White Bassw'd,	1/8 "	5 "	9c.	Satin Wood,	1/8 "	25 "	36c.
" "	3/16 "	6 "	11c.	Tulip,	1/8 "	35 "	47c.
" "	1/4 "	7 "	14c.	Ebony,	1/8 "	45 "	68c

☞ Woods over 12 or 14 inches square cannot be sent by mail, and not less than two pieces the same length and width can be sent, as it is necessary to cross the grain to prevent splitting.

PATTERN OF LORD'S PRAYER AND WINDOW GARDEN SCENE.

Lord's Prayer.............. 25c.
Window Garden Scene, all complete.............. 50c.
" " " any pattern separate, each.............. 15c.

PRICES OF PRINTERS' MATERIALS FOR MAKING WOOD TYPE:

Prepared Holly, 1-16 and ⅛ thick, per sq. ft.............. 10c.
" Pine, proper thickness to make Holly type high, per sq. ft.............. 10c.
Giant Glue, for gluing Holly to Pine, per bottle.............. 25c.
15 Full Set of Alphabets, printed for patterns.............. 1.00

Add 10 cents extra to postage on Canadian orders. Packages to Canada restricted to 8 ounces. Estimate an ounce for every CENT WE ASK FOR POSTAGE AND YOU HAVE THE WEIGHT OF OUR GOODS. Fancy woods safe and cheap by freight—add 10 cents for crate on every 40 feet or under.

SHIPMAN'S MANUAL OF 87 ASSORTED PATTERNS, 50 CTS.

A selection of desirable designs, Six Alphabets, also an Elaborate Treatise on Bracket Sawing, Inlaying, Preparation of Woods, Polishing, &c.

PRICES OF PARTS OF PRIZE HOLLY SCROLL SAW AND LATHE.

In ordering, say "for Prize Holly;" or, what is better, take picture of one and mark the part and place with fine pen and ink, and mail us.

		BY MAIL.			BY MAIL.
Straining rod	$ .15	$ .15	Collar for head block	$ .15	$ .17
Upper wooden arm	.10	.15	Holly driving Disc	.20	.25
Lower wooden arm	.10	.15	Head block pulley	.25	.35
Clamps, per pair	.25	.25	Head block, all complete	1.00	1.35
Emery wheel	.50	.65	Head block, frame only	.25	.35
Stud pin, for drive wheel	.25	.25	Tail block, " "	.25	.35
4-inch rest, for turning	.15	.17	Tail block, all complete	1.00	1.20
Rest socket, for rest	.20	.25	Pitman, or treadle connecting rod	.10	.12
Hanger spindle	.10	.15	Turning tools, each	.25	.25
Head block spindle	.30	.35	Belts, each	.20	.25
Tail block spindle	.25	.30	Bolts and set screws, each	.05	.05
Spur centre	.15	.17	Tail screw handle	.15	.20
Spindle for buzz saw and dovetailing attachment	.30	.35	Hanger	.25	.40

PARTS THAT MUST GO BY EXPRESS

To ensure the delivery of these heavier parts at a great distance from here, send order through your local Express Agent, or have him write Rochester Express Agent, that your order will be received by you, as Express companies require guarantee on goods not of great value, going a far distance. It will cost no more and you will be certain.

Polished tilting table	$ .30	Lathe bed		$ .75
Arm piece	.50	Driving Wheel		.75
Right, left or back leg, each	.35	Wrought iron treadle bar, with collars		.40
Treadle	.35			

PRICES OF PARTS OF PRIZE DEMAS SCROLL SAW AND LATHE.

Be certain and use the words "for Prize Demas," in ordering these parts.

		BY MAIL.			BY MAIL.
Head block spindle	$ .30	$ .35	Demas stud pin, for drive wheel	$ .25	$ .25
" " pulley	.25	.35	4-inch rest	.15	.17
Tail block screw	.25	.30	12-inch rest	.30	.38
" " handle	.15	.20	Rest sockets, per pair	.40	.50
Spur centre	.15	.17	Pitman connecting treadle	.10	.12
Collar for head block spindle	.15	.17	Turning tools, each	.25	.25
Head block, frame only	.25	.35	Screw driver	.20	.25
Tail " " "	.25	.35	Belts, each	.20	.25
Wooden arms, each	.10	.15	Demas driving Disc	.25	.30
Emery wheel	.50	.65	Under plate and screws, upper arms	.15	.15
Saw fasteners, per pair	.25	.25			
Straining rod	.15	.15			
Bolts and set screws, each	.05	.05			

PARTS THAT MUST BE SENT BY EXPRESS.

Lathe bed	$1.00	Wrought iron treadle bar, with collars	$ .50
Polished tilting table	.50	Wire brace	.15
Treadle	.40	Head block, complete	1.00
Drive wheel	1.00	Tail " "	1.00
Arm piece	.75		
Side legs, each	.65		

Parts of Nos. 4 and 5 Demas, for Mechanics—By Freight or Express.

Head Block, complete	$3.50	Lathe Bed for No. 4	2.50
Tail " "	3.50	" " " 5	1.50
Drive Wheel, "	4.00	Steel Shaft for drive wheel	1.00
Treadle Rod, "	.75	Boxes for same, 25c, by mail	.50
Treadle Wood Work, complete	.75	Crates for either one or two legs or a number of parts additional	.50
Saw Table	.75		
Legs, each	2.00	Crates for either or both Lathes, alone	.25
Iron Arm Piece	1.50		

PARTS MAILABLE.		By Mail	PARTS MAILABLE.		By Mail
Wooden Saw Arms, each	$ .20	$ .25	Binder Wheels for Head Blocks, Tail Blocks and Rests, and Saw Arms	$ .40	$ .60
Saw Clamps	.50	.50			
Cranks	.25	.80	Small Balance Wheel for 4 and 5	.50	.79
Rests	.25	.40	Face Plate	.50	.70
Rest Sockets	.10	.70	Thumb Screw	.05	.05
Tail Block Hand Wheel	.50	.70	Bolts and Set Screws, each	.10	.10
Shieve Wheel	.50	.60			

Rosette Chuck for NO. 4 Lathe, 75 cents; by mail, 80 cents.
Prize Holly Buzz Saw Attachment, by Mail, $1.30.
Prize Demas, " " 2.00.
Purchasers not having a Lathe Head Block, should ask for "BUZZ SAW SPINDLE," 30 cts. extra.